GEOLOGISTS' ASSOCIATI

GW01401414

ONNY VALLEY, SHROPSHIRE

GEOLOGY TEACHING TRAIL

by PETER TOGHILL
School of Continuing Studies
University of Birmingham

Edited by J. T. Greensmith

PREFACE

The type area and type sections for many of the stages of the Caradoc Series of the Ordovician System occur in south Shropshire. The Onny Valley, to the east of the Church Stretton Fault Complex, provides a classic and famous section through the Series and along it are exposed all the characteristic rock varieties containing their well-known shelly faunas.

One of the fundamental arguments for the erection of the Ordovician System was based on evidence from the Onny Valley, where the major unconformity (cover photograph) between Murchison's Upper and Lower Silurian was first identified; the latter is now essentially the Ordovician.

The 1958 Geologists' Association Guide to *South Shropshire No. 27* (revised in 1968) included the Onny Valley as one of the itineraries. This present guide is a more comprehensive account, visiting exposures which show all the typical rock formations. Hammering of rock faces is not allowed, but specimens and fossils can be collected from loose scree. You are asked not to divert from the stated route.

The trail guide is meant to appeal to 'A' level students and undergraduates, as well as adult education students and amateur geologists. The general public will find something to interest them too, as the trail follows a beautiful wooded valley, with plenty of wildlife to see. Please respect the countryside.

In valleys of springs of rivers,
By Onny Teme and Clun,
The county for easy livers,
The quietest under the sun.

A Shropshire Lad, A. E. Houseman

THE ORDOVICIAN GEOLOGY OF SHROPSHIRE

ORDOVICIAN System/Period	Ashgill Caradoc Llandeilo Llanvirn Arenig Tremadoc	Series/Epoch

No other area of comparable size in the British Isles shows such a variety of geology as does Shropshire, with all geological periods (except the Cretaceous and Tertiary) represented by sedimentary rocks, and many also by igneous rocks, including the Tertiary. The sequence has been affected by five periods of earth movements associated with major orogenies and these have resulted in numerous unconformities, folds, faults and episodes of igneous activity.

The Ordovician System, the sequence of rocks laid down during the Ordovician Period, is divided into six Series representing the six Epochs of the Period.

The Tremadoc Series has in the past been considered as late Cambrian, and still is by some geologists. Both east and west of the Pontesford-Linley Fault the Series is represented by the Shineton Shales, grey-green shales up to 1000m thick, which in the Shelve area, west of the Pontesford-Linley Fault, pass up into the Arenig Series Stiperstones Quartzite with little break in the sedimentary sequence.

Rocks of the Ordovician Period, 510-435 million years old, crop out in five main areas in Shopshire and are well known for the fact that they exhibit differences in sequence west and east of the Pontesford-Linley and Church Stretton Fault Complexes (Figure 1).

They crop out in the Breidden Hills, and west of Oswestry where the sequence is part of the larger Ordovician area of the Berwyn Dome and in these areas the rocks range in age from the Llandeilo through into the Caradoc epoch. In the Berwyns, Ashgill rocks are present.

In the classic western area of the Shelve Inlier an almost complete sequence occurs passing upwards from the Tremadoc Series through the Arenig, Llanvirn, Llandeilo and well into the Caradoc Series, with a variety of sedimentary rocks, extrusive lavas and volcanic ashes. Pauses in deposition are recognised in the Tremadoc Series.

In the eastern type Caradoc area, and also in the small but critical area around Pontesford just east of the Pontesford-Linley Fault, rocks of the Caradoc Series occur resting unconformably upon Tremadoc Series Shineton Shales, especially noticeable around Hoar Edge, some 6km northeast of Church Stretton. Arenig, Llanvirn and Llandeilo rocks were not laid down here because the area was then a landmass created by a retreat (regression) of the sea to areas west of the Pontesford-Linley Fault at the end of the Tremadoc epoch. The sea did not spread (transgress) across this landmass again until the beginning of the Caradoc epoch.

The Caradoc area lies east of the Church Stretton Fault (Figure 1), although the actual boundary between its characteristic Ordovician succession and those of the Shelve and more westerly areas is the Pontesford-Linley Fault.

The Caradoc succession in this type area consists of shallow water, marine sandstones, shales and thin limestones rich in animals living on the floor of the sea (benthos). Volcanic rocks are absent. This type of succession is often referred to as a 'shelly facies', in contrast with a 'graptolitic facies' represented by thicker successions of deeper water sedimentary rocks further to the west in Wales, but also in some of the lower Ordovician of the Shelve area. Ashgill rocks are absent as they are throughout most of the county except in the far northwest. From this it is deduced that the Ashgill epoch was a time of earth movements in Shropshire, when no marine sedimentation took place because most of the county was above sea-level, with a north-south shoreline running from Oswestry southwards towards Welshpool.

These earth movements relate to the late Ordovician collision between a northern Baltic and southern Avalonian Plate during the closing of the Iapetus Ocean, which separated England, Wales and southern Ireland from Scotland and Northern Ireland. Southern Britain was probably 35° south of the equator during these times. The movements are usually called Taconian, but the author (1990) has now suggested the term Shelveian movements or event, named after the folding and faulting of this age in the Shelve area. The retreat of the sea westwards at the end of the Caradoc and during the Ashgill is partly explained by these movements, but it might also have been significantly affected by lowered sea-levels on a worldwide scale, due to the growth of extensive ice sheets on an ancient continental landmass, known as Gondwanaland, and centred on what is present day Morocco.

In Shropshire widespread contemporaneous intrusive igneous activity produced such well-known features as the Corndon dolerite phacolith (a lens-shaped body occupying the crest of a fold) of the Shelve Inlier and the Criggion dolerite in the Breiddens (both actually across the border in Wales). The lead-zinc-barytes mineralisation in the Shelve area occurs in Ordovician rocks, but is probably of early Carboniferous age.

THE GEOLOGY OF THE TYPE CARADOC AREA

Caradoc Series	Onnian Stage
	Actonian Stage
	Marshbrook Stage
	Woolstonian Stage
	Longvillian Stage
	Soudleyan Stage
	Harnagian Stage
	Costonian Stage

Of the six Series of the Ordovician System, the Caradoc Series has its type area in Shropshire, east of the Church Stretton Fault Complex. A type area is one where the rocks of a System or Series are best studied and well exposed, and can be subdivided into various Stages each with their diagnostic fossils, which allow for accurate correlation all over the world.

The Caradoc Series has been divided into eight Stages. The various stage names shown above are derived from places within the area.

The type Caradoc area is quite a large strip of Ordovician rocks trending northeast – southwest for 31km from Harnage in the north to Coston in the south, with a maximum width of 3km (Figure 1). The strip is split into two by an outcrop of Precambrian and Cambrian rocks east of Church Stretton.

The rocks rest with marked unconformity on older rocks ranging in age from Precambrian (Longmyndian and Uriconian) through to earliest Ordovician (Tremadoc Series) Shineton Shales. They dip southeast away from the older rocks due to Shelveian movements and steep dips near to the Church Stretton Fault Complex become less steep further east.

The type Caradoc rock sequence

The table below lists the rock units of the area. The sequence of rock formations used here is the classification adopted by the British Geological Survey (Greig *et al*, 1968) but others, notably Dean (1958, 1964), have used other terms as will be explained below.

In the far south of the area around the Onny Valley, Dean (1958) has used the terms Coston Beds, Smeathen Wood Beds and Glenburrell Beds in place of Hoar Edge Grits and Harnage Shales, since the lithologies are rather different to those occurring further north. Dean also prefers the term Horderley Sandstone in the far south to include the Chatwall Flags and Chatwall Sandstone which are diachronous units. This latter term means that the age of the rock units is not the same over the whole area, as different sedimentary environments arrived at different times as sea spread over the area.

Just to the east of Church Stretton, where the Precambrian/Cambrian outcrop splits the Ordovician outcrop into two (Figure 1), the basal Hoar Edge Grit Formation is missing around Hope Bowdler, and the Harnage Shales rest with a marked unconformity on Precambrian volcanics. This is seen at the classic roadside exposure of Hope Bowdler (Toghill, 1990, figure 30). Neptunean sediment-filled dykes of Harnagian age occur on Hazler Hill within the Uriconian volcanics just to the west of Hope Bowdler. As well as this

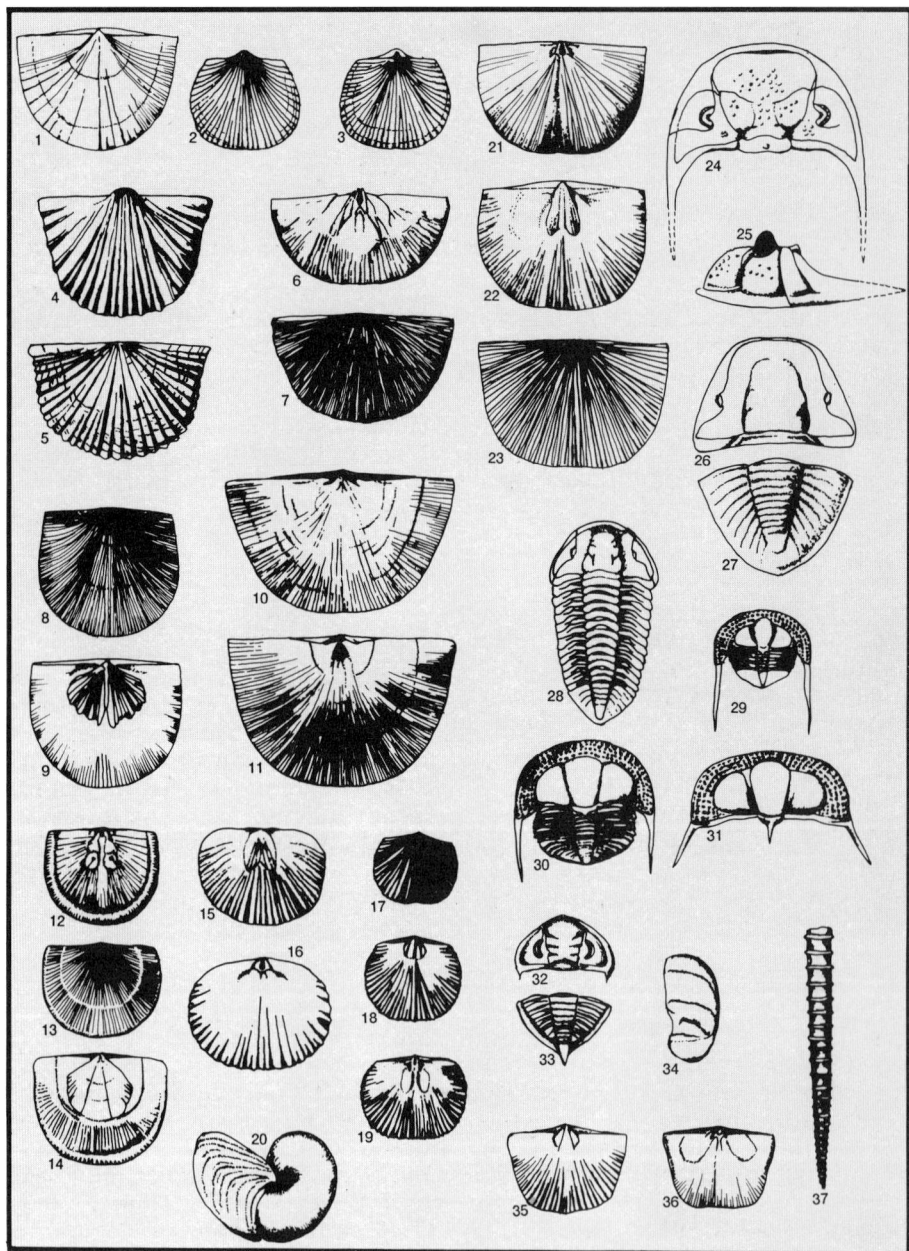

Figure 2

Stages	Formations
Onnian	Onny Shales
Actonian	Acton Scott Group
Marshbrookian	Cheney Longville Flags
Woolstonian	*alternata* Limestone
Longvillian	Chatwall Sandstone
Soudleyan	Chatwall Flags
Harnagian	Harnage Shales
Costonian	Hoar Edge Grit

Fig. 2. (Opposite) Diagnostic fossils from the Caradoc Series rocks in the Onny Valley.

1. *Kjaerina bipartita*, x0.5; 2, 3. *Onniella roeggeri*, x1.25;
4, 5. *Nicollella actoniae*, x1; 6, 7. *Sowerbyella sericea*, x1;
8, 9. *Macrocoelia expansa*, x1; 10, 11. *Strophomena grandis*, x0.5;
12, 13, 14, *Heterorthis alternata*, x0.5; 15, 16. *Dinorthis flabellulum*, x0.75;
17, 18, 19. *Dalmanella horderleyensis*, x0.75; 20. *Sinuites*, x0.7;
21, 22, 23. *Harknessella vespertilio*, x0.7; 24, 25. *Chasmops extensa*, x0.5;
26,27. *Brongniartella bisulcata*, x0.5; 28. *Flexicalymene caractaci*, x0.7;
29. *Onnia superba*, x0.7; 30. *Broeggerolithus broeggeri*, x1.25;
32, 33. *Kloucekia apiculata*, x0.75; 34. *Tallinella scripta*, x5;
35, 36. *Reuschella horderleyensis*, x0.7; 37. *Tentaculites*, x1.25.

variation to the normal sequence, the Onny Shales, the youngest of the Caradoc rocks only occur around the Onny Valley and Acton Scott areas. Elsewhere they are cut-out by the overlying basal Silurian rocks.

Hoar Edge Grit

The basal formation of the Caradoc Series (absent in the vicinity of Hope Bowdler) varies in thickness from 0-120 metres. It rests with a marked unconformity in the north on Tremadoc Series (earliest Ordovician) Shineton Shales, where it forms the conspicuous escarpment of Hoar Edge, east of The Lawley, and here it reaches its maximum thickness. Around Hoar Edge it comprises yellow-brown sandstones and pebbly sandstones, with wind-faceted pebbles (dreikanters) suggesting the proximity of an arid coastline to the east. Further north calcareous sandstones and shelly layers appear. In the south calcareous sandstones, unconformable on Precambrian rocks, occur in the Onny Valley often with pebbly layers, and with a basal conglomerate at Coston to the west of Craven Arms. Brachiopods present include *Harknessella* and *Dinorthis*, and the trilobite *Costonia* is characteristic.

Harnage Shales

100-300 metres of micaceous grey-green silty mudstones and shales, often orange-stained, and with occasional thin sandstones, is the typical sequence in the north of the area and around Hope Bowdler. In the south, in the Onny Valley, the sequence contains grey-green blocky mudstones and siltstones, and less shale, and here Dean uses the terms Smeathen Wood Beds and Glenburrell Beds.

The outcrop often forms low ground in the valley between Hoar Edge and the Chatwall Sandstone escarpment of Yell Bank just to the east at Chatwall. Around Hope Bowdler the ostracod *Tallinnella* is common and elsewhere the trilobite *Salterolithus* is characteristic, also the brachiopods *Salopia* and *Smeathenella* occur. The Harnage Shales usually succeed the Hoar Edge Grit with quite a sharp change in lithology from sandstones to shales, as in the Onny Valley, but around Hope Bowdler the Harnage Shales is the basal Caradoc formation and comprise shales with a thin basal conglomerate resting on Uriconian volcanics.

Chatwall Flags

The Harnage Shales grade up into 30-100 metres of flaggy, fine grained greenish brown sandstones with shale bands, the Chatwall Flags. They form the lower part of the Horderley Sandstone of Dean (1958) and topographically form the lower slopes of the escarpments formed by the Chatwall Sandstone. They contain the brachiopods *Macrocoelia* and *Sowerbyella* as well as circular ossicles of the fragmented crinoid stems of *Balacrinus*. The trilobite *Broeggerolithus* also occurs.

Chatwall Sandstone

This well known sandstone forms conspicuous escarpments at Chatwall (Yell Bank), at Soudley, and at Briar Edge between Marshbrook and the Onny Valley. It consists of 40-160 metres of conspicuously banded brown, green and purple sandstones, often massively bedded. Current-bedding, cross-stratification and hummocky bedding are often clearly visible and are considered the result of storm disturbance in shallow marine environments (Brenchley & Newall, 1982; Hurst, 1979.) Trilobites include *Broeggerolithus* and *Kloucekia* and brachiopods such as *Sowerbyella*, *Bancroftina*, *Dalmanella* and *Howellites* occur, as well as the gastropod *Sinuites*.

8

This is the well-known building stone of the area so well seen in villages such as Hope Bowdler, Cheney Longville and Cardington. In the Onny Valley it is 150 metres thick and was quarried extensively in the past at Horderley Quarry. In this area many authors, including Dean (1958), prefer to use the name Horderley Sandstone. Further north, at Soudley Quarry near Hope Bowdler, it is often called the Soudley Sandstone and is only 45 metres thick.

alternata Limestone

Separating the Chatwall Sandstone from the overlying Cheney Longville Flags are 20-30 metres of lenticular shelly limestones interbedded with green micaceous flags and shales. The whole sequence is highly fossiliferous and the limestone lenticles occurring throughout represent fossil shell banks between 0.25 and 0.75 metres thick. They contain an abundance of the brachiopod *Heterorthis alternata*, hence the name of the formation, as well as *Sowerbyella*. Hurst (1979) has shown the formation to be the product of migrating marine bars and storm deposition.

Cheney Longville Flags

80-240 metres of greenish grey flaggy sandstones and shales with thin shelly limestones in which cross-stratification and hummocky bedding are the result of storm influenced shallow water sand deposition in the form of lobes and sand bars (Hurst,

1979; Brenchley & Newall, 1982). They are well exposed around Cheney Longville and the rest of the type Caradoc area and contain many brachiopods, particularly *Kjaerina* and *Dalmanella*, trilobites, including *Chasmops*, *Brongniartella*, *Broeggerolithus* and *Flexicalymene*. *Tentaculites*, a scaphopod mollusc, is often abundant and current-orientated on certain bedding planes.

Acton Scott Group

This Group consists of 60-80 metres of pale brown and yellow brown mudstones and sandstones. Around Acton Scott a development of pale calcareous sandstones, the Acton Scott Limestones, occurs in the middle of the formation and is a local building stone. Trilobites include *Chasmops* and *Platylichas* and the brachiopods include *Nicollella*, *Strophomena*, *Onniella* and *Reuschella*.

Onny Shales

These, the youngest rocks of the Caradoc Series, belong to the Onnian stage and comprise 0-120 metres of grey micaceous siltstones which, towards the top of the sequence, show a typical orange weathering effect. They are well known for trinucleid trilobites of the genus *Onnia* and the brachiopod *Onniella*. The rocks only outcrop in the Onny Valley and around Acton Scott, being cut-out north and south of these localities by the overstep of the overlying unconformable Silurian rocks.

THE GEOLOGICAL TRAIL THROUGH THE ORDOVICIAN (CARADOC SERIES) ROCKS OF THE ONNY VALLEY

The 1.75km long trail along the Onny Valley demonstrates in detail the varieties and geological structure of the Caradoc rocks (formations) in their type area (Figures 3 and 4).

Here, all the various formations can be examined in ascending order travelling from northwest to southeast, from the base of the Hoar Edge Grit, which rests unconformably on Precambrian Western Longmyndian rocks, through to the Onny Shales. At the famous unconformity exposure on the north bank of the River Onny, south of Wistanstow, the Onny Shales are overlain unconformably by the Silurian Hughley (Purple) Shales of Late Llandovery age. The rocks strike NE – SW and are almost vertical in the proximity of the Church Stretton Fault Complex, but dip more gently southeast away from the fault (Figure 3).

In ascending the geological sequence the trail follows the line of the disused Bishop's Castle railway and the south bank of the River Onny. *On no account should the River Onny be crossed to its northern bank as this is outside the ownership of Cheney Longville Estates.*

Many of the rocks are fossiliferous but *hammering of rock faces is not allowed*, although fossils and specimens may be collected from local scree. The diagnostic fossils of each rock formation are listed in the previous pages and shown in Figure 2.

The trail starts at the base of the sequence and parties having followed the whole trail can retrace their steps to the start. However, 400m east of the last locality (8), the unconformity exposure, a footbridge crosses the River Onny and leads north to Crossway (SO 429852) on the A489, a distance of 150m, and here parties can be picked up.

Location of the trail and parking

Leave the A49 1.5km north of Craven Arms and travel west on a minor road signed to Cheney Longville. The village is reached in 1km and then the route continues through the village NW on a narrow but surfaced road towards Longville Common and Edgton. After 1km turn sharp right downhill on a narrow surfaced road at SO 412855 which leads north down to the Onny Valley. This is a No Through Road and after 400m it becomes a grassy track. At this point (SO 413858) (Figure 4) cars and vans can be parked on the wide grass verges. Please do not block the road. There is no access through either of the two gates on the left near to this point, and these entrances should be left clear for farm vehicles. Do not drive any further north from here towards the old railway bridge.

Note that this route from the A49 is impossible for coaches. Coach parties can be dropped at Crossway (SO 429852) on the A489 (see above) and the trail followed in reverse.

Locality 1: Quarry in Longmyndian and Hoar Edge Grit on south bank of the River Onny

Grid ref: SO 41158614

Proceed north on foot along the grass track across the old railway bridge (recently restored by the Bishop's Castle Railway Society) and then northwest along the south bank of the River Onny for 250m to a gate with a stile which gives access to a well-known quarry on the south bank of the River Onny, opposite the Round House.

This quarry exposes 22m of Hoar Edge Grit (Coston Beds of Dean, 1958) dipping southeast at 65° to 70°. The Grit unconformably overlies Western Longmyndian sandstones and is succeeded by the basal beds of the Harnage Shales (Smeathan Wood Beds).

The Hoar Edge Grit comprises grey, brown and yellow-brown, medium grained calcareous sandstones, often with conspicuous lenticular solution hollows and cavernous weathering. Thin, grey crystalline limestones occur as well as coarser quartz sandstones, with pebble beds at certain horizons. Pebbles are usually less than 1cm in diameter but can be up to 5cm. Brachiopod casts are present in the coarser sandstones.

Parts of the sequence are massively bedded, whereas others are more thinly bedded and show current-bedding. The steep faces of the quarry (*approach with care*) are major NW – SE joint planes. The lenticular solution hollows noticeable throughout are probably caused by lime-rich layers weathering out, although Whittard (1952) suggested that they were caused by partial or complete weathering out of polyzoan colonies.

Details of the Precambrian and Ordovician-Precambrian boundary

At the western end of the quarry a flight of steps leads to the junction between the Hoar Edge Grit and the underlying Precambrian Western Longmyndian sandstones. Dean (1964) considers this junction to be a fault which has cut-out the basal conglomerates of the Hoar Edge Grit (Coston Beds) outcropping further to the south. The British Geological Survey consider the junction to be an unconformity, pointing out (Greig *et al.*, 1968 p.119) that the disappearance of the basal conglomerates in this area is probably a depositional feature, although some minor faulting and

overthrusting has probably taken place at the boundary during the late Ordovician (Shelveian) folding. In this guide the unconformable nature of this junction is accepted and the steep dips are considered to be due to late Ordovician folding and movements along branches (F1, F2 and F3) of the Church Stretton Fault Complex which are close-by to the east and west.

The Precambrian Western Longmyndian rocks are exposed at the far western corner of the quarry to the right of the steps. They consist of blocky, well-jointed, medium grained brown sandstones, dipping 70° SE. Only 1m is exposed with two good joint directions, one vertical running NW-SE, and the other inclined at 25° NW. The first set is parallel to that in the overlying Hoar Edge Grit which forms the main quarry faces.

Please do not hammer the Longmyndian outcrop as the amount of rock exposed is very small and this is an important exposure.

The hard Longmyndian sandstones pass up (to the east) into 1.8m of deeply weathered, purple-brown sands containing lumps of weathered Longmyndian sandstone. A 2 to 3cm thick layer of lighter coloured sand, clayey in places, occurs at the top of the sequence just below the first obvious beds of Hoar Edge Grit. *Do not hammer.*

There are two possible explanations for the formation of the weathered sands. They could be explained by deep sub-aerial weathering of the Longmyndian during some part of the Cambrian and early Ordovician, when the area was a land mass and prior to the spreading of the sea from the west during Caradoc times with the formation of the Hoar Edge Grit. If so, the weathering indicates the scale of the unconformity.

On the other hand, it is possible

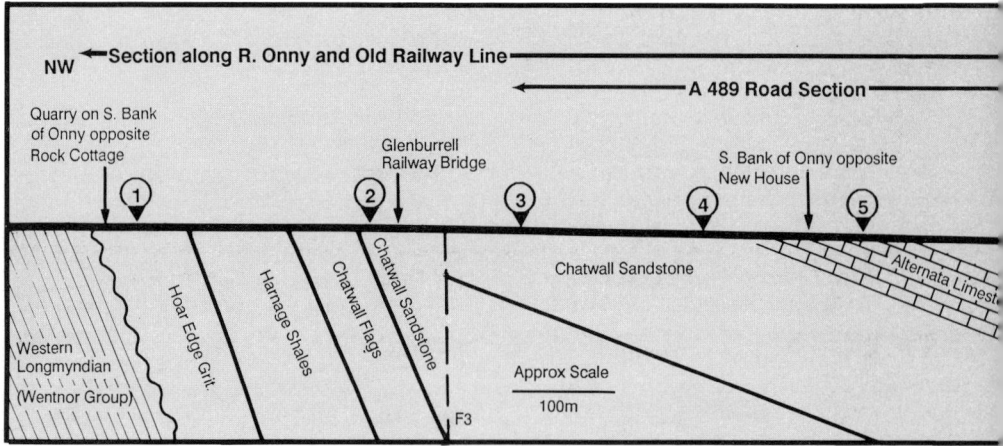

Fig. 3. Section along the Onny Valley Trail.

Fig. 4. Plan of the Onny Valley Geological Trail.

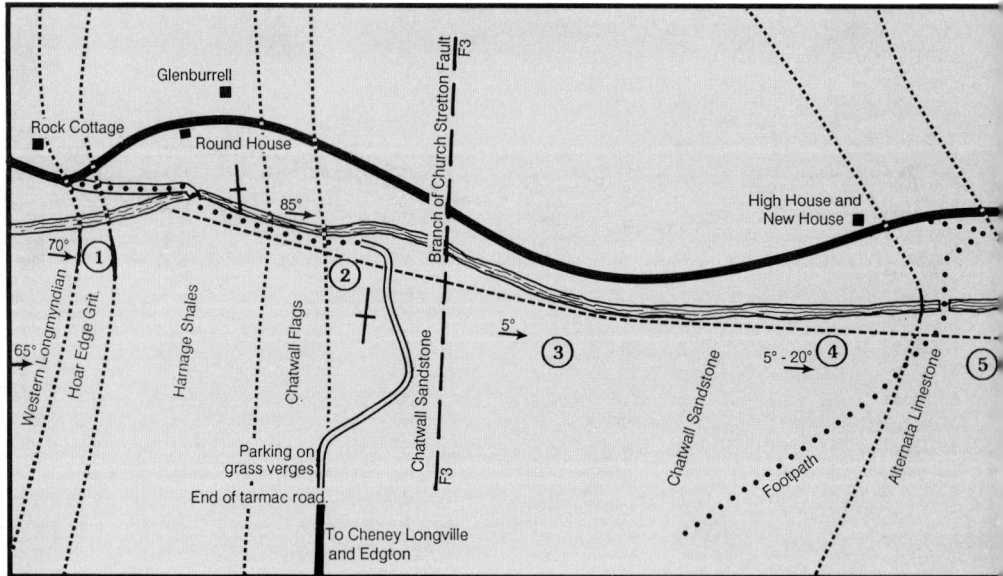

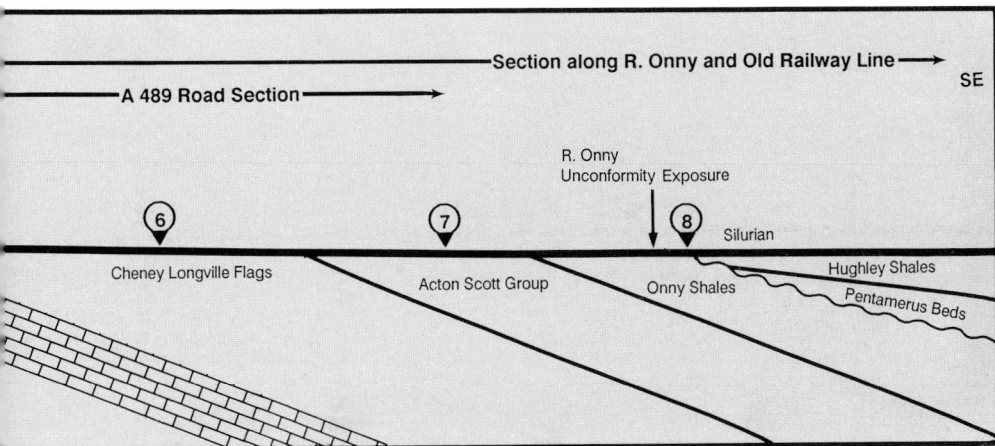

Section along R. Onny and Old Railway Line →

SE

A 489 Road Section →

R. Onny
Unconformity Exposure

⑥

⑦

⑧

Silurian

Cheney Longville Flags

Acton Scott Group

Onny Shales

Hughley Shales

Pentamerus Beds

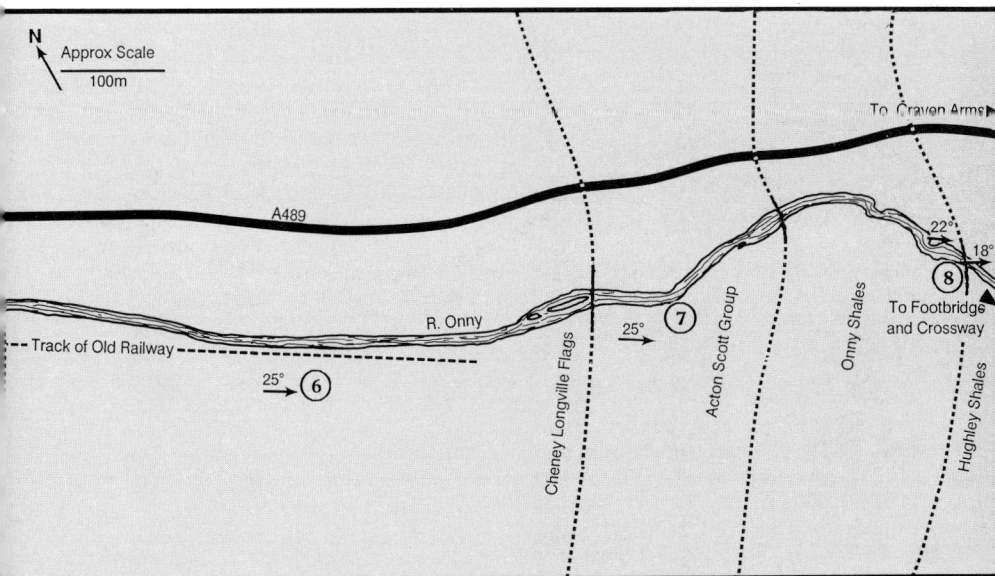

N

Approx Scale
100m

To Craven Arms

A489

22°

18°

⑧

Track of Old Railway

R. Onny

25°

⑦

To Footbridge
and Crossway

25°

⑥

Cheney Longville Flags

Acton Scott Group

Onny Shales

Hughley Shales

13

that the soft incompetent sands at the top of Longmyndian were caused by crushing during Shelveian faulting and overthrusting along the boundary at the end of the Ordovician, and were not a result of pre-Caradoc weathering.

The 2 to 3cm clayey sand layer at the top of the Longmyndian is followed by the basal beds of the Hoar Edge Grit, which are weathered orange-brown sandstones, with some purple colour derived from the underlying Longmyndian sands. They are soft in places and contorted by minor faulting, but within 0.3m of the base they pass up into hard, bedded grey-brown calcareous sandstones. Some overthrusting is visible to the SE along an obvious joint plane in the Hoar Edge Grit dipping 40 NNW, two thirds of the way up the quarry face, and the Ordovician-Precambrian unconformity is slightly contorted as shown by the curvature of the light coloured sand layer at the top of the Longmyndian. The beds above this joint plane in the Hoar Edge Grit show more fracturing and contortion than those below.

Details of the Hoar Edge Grit

The quarry is divided into western and eastern parts showing lower and higher divisions of the Hoar Edge Grit. A major joint plane dipping 40° N forms a link between the two parts whose steep faces are NW-SE trending major joints. *Approach both faces with care.*

The western face shows the lowest beds to be 3m of well bedded, pale brown, medium grained calcareous sandstones with some current-bedding. One obvious softer sandstone band 0.5m from the top carries brachiopod casts. There are some weathered-back calcareous nodule horizons and thin grey crystalline limestones are also present.

These basal beds dip 65° SE and pass up into 4.3m of more thinly bedded, medium grained, grey-brown calcareous sandstones with current-bedding and thin crystalline limestones. Obvious lenticular solution hollows occur elongated parallel to the bedding, possibly where polyzoan colonies, or lime-rich layers, have partially or completely weathered out.

The eastern part of the quarry continues the sequence, dipping 75° ESE, with 6.5m of massively bedded, pale grey calcareous sandstones and thin grey crystalline limestones with few bedding planes. The massive beds pass up into 5m of more obviously thinly bedded and current-bedded, pale grey-brown calcareous sandstones with large lenticular solution hollows.

They are succeeded by 2.3m of massively bedded, grey-brown medium grained quartz sandstones, followed by 1.75m of orange-brown quartz sandstones showing current-bedding and with some shaley bands. The highest 0.8m of this sequence contains many brachiopod casts and larger solution hollows. The quartz grains are well rounded and in conglomerate horizons occur well rounded quartz pebbles up to 5cm in diameter, though generally smaller.

There is then a rapid change to steeply dipping olive-green shales in the eastern corner of the quarry. These are the Harnage Shales (Smeathen Wood Beds of Dean, 1958). They are contorted due to soil creep but about 2m are exposed. This is the only permanent exposure of Harnage Shales along the trail. *Please do not hammer.* However, at times of low water in the River Onny there are exposures in the stream bed of up to 75m of blocky grey-green micaceous mudstones with an almost vertical dip and striking NE-SW, between localities 1 and 2. *On no account cross the River Onny to the north bank.*

14

Locality 2: Old Railway Cutting west and east of Glenburrell Rail Bridge, exposing Chatwall Sandstone.

Grid ref: SO 41318604

Return east from locality 1 towards the old railway bridge, now restored by the Bishop's Castle Railway Society, and gain obvious access to the old railway line running towards the bridge.

On the south side of the track is a cutting with clear exposures of Chatwall Sandstone in which the beds are all nearly vertical. A 70m section is exposed before the bridge is reached and this lies within the lower part of the Chatwall Sandstone. The base of the sandstone and the underlying Chatwall Flags are not exposed.

The first 50m of the sequence, dipping about 85° SE, consists of fine to medium grained sandstones, green-brown, occasionally banded purple, green and brown, and often coated with haematite. They are massively bedded in places but with conspicuous olive green shale bands up to 20cm in thickness. Hummocky tops to the sandstones and current-bedding are indictive of storm-influenced deposition in shallow marine environment, near to the shoreline (Brenchley & Newall, 1982). This type of bedding is noticeable near to a 20cm shale band at the beginning of the exposures, and also after walking about 50m.

From this point as far as the bridge the beds continue nearly vertical, but are heavily fractured due to the proximity of the most easterly (F3) branch of the Church Stretton Fault Complex. East of the bridge vertical sandstones continue for a further 50m and then there is an exposure gap of 100m before almost horizontal Chatwall Sandstone is reached at locality 3. The F3 branch of the Church Stretton Fault Complex has been crossed in this exposure gap, as shown on the British Geological Survey maps.

Locality 3: Small quarry in typical Chatwall Sandstone, 60m east of Glenburrell railway bridge.

Grid ref: SO 41478587

Continue east along the old railway line for 120m from the bridge (or 70m from the end of the cutting) and then branch off slightly to the right, down the embankment, and continue SE at a slight angle to the embankment for about 25m following an old fence with boggy ground below the embankment on the left. Then go uphill for 20m to an obvious small quarry in Chatwall Sandstone. The exposure shows banded purple, green and brown sandstones typical of building stones within the middle of the Chatwall Sandstone sequence (Horderley Sandstone of Dean, 1958). 6m of massively bedded, medium grained sandstones are exposed with flaggy layers and thin calcareous lenses carrying abundant shells of brachiopods and other fossils. Cross-bedding and hummocky bedding is visible at this quarry, indicative of storm influenced deposition, as at locality 2. *Do not hammer this locality.* Fossils can be collected from locality 4.

The dip here is very gentle, only around 5° SE, confirming that the F3 branch of the Church Stretton Fault Complex has been passed over between localities 2 and 3. Two almost vertical sets of joints at 120° to each other from the quarry faces.

Locality 4: Old railway cutting showing highest Chatwall Sandstone.

Grid ref: SO 41708574

Return to the old railway track and continue east for 300m towards where a gate crosses the track. Here, exposures in the highest Chatwall Sandstone seen along the trail crop out for 50m west of the gate on the south side of the track.

Approximately 6m are exposed with dips varying from 5° to 20°SE.

The sandstones are greenish brown, well bedded and with numerous shell bands up to 10cm in thickness. *Do not hammer the face.* The dip decreases from west to east and this may be a reflection of variations in the deposition dip due to the prevalence of storm environments with migrating marine sand bars in those times. (Hurst, 1979).

Within the 50m long outcrop ample scree provides numerous fossils, including brachiopods, bivalves, gastropods and trilobites. Further west the outcrops continue higher up in dense woodland and, again, screes provide abundant fossils.

Locality 5: Old railway cutting showing *alternata* Limestone.

Grid ref: SO 41808568

Continue east from locality 5 over the stile by the gate and after 100m a low cutting on the south side of the old railway track, opposite a footbridge over the River Onny, shows scattered exposures in greenish brown micaceous sandstones and flagstones and lenticular grey-green shelly limestones. The latter are packed with the brachiopod *Heterorthis alternata*, which is why this formation is called the *alternata* Limestone. Hurst (1979) considers that the deposits reflect the influence of near and distant storms and are associated with migrating marine bars.

Exposures are poor. *Please do not hammer the outcrops.* The thickness is difficult to estimate but 25m has been mapped in this area.

Locality 6: Old river cliff on south side of old railway track showing exposures in Cheney Longville Flags.

Grid ref: SO 42078547

Continue along the old railway track from locality 5 for 350m and, 50m before a gate is reached, bear right off the embankment to obvious exposures above a boggy pond. Avoid the cliff directly above the pond as the ground is very marshy, but a track above the cliff leading SE has exposures on the south side showing 20m of Cheney Longville Flags dipping 25°SE.

These comprise thinly bedded, grey-green, micaceous, fine grained sandstones, siltstones and mudstones, hence the term "flags", Thin shell bands occur and some bedding planes are crowded with the straight-shelled, ribbed scaphopod *Tentaculites.*

Some of the micaceous siltstones are very finely laminated and colour banded grey and greenish grey. They show current-bedding and channel structures cutting down into the underlying mudstones and siltstones. Many of the flaggy sandstones are not so finely laminated and have flat bedding planes, whereas others show conspicuous hummocky bedding on their top surfaces and these are often followed by blocky unlaminated mudstones. The current and hummocky bedding are features now considered to be indicative of storm wave action in shallow marine environments associated with migrating sand bars (Hurst, 1979; Brenchley & Newall, 1982). *Please do not hammer this locality. Collect from the scree.*

Locality 7: River side exposure on south bank of River Onny in Acton Scott Group.

Grid ref: SO 42388538

Return to the railway track from locality 6 and continue east for 50m to a gate and climb the stile to the right of it. After a further 70m, a fence crosses the old railway track and can be crossed by a stile. Then bear immediately left and avoiding the old railway bridge pass through a field towards the River Onny. Follow the field fence above the river for 160m eastwards until a stile gives access to the river bank.

Go down to the river and downstream (east) for 15m to a small cliff above the river about 5m high. The top 2.5m of the cliff shows coarse river gravels and these overlie 2.5m of the Acton Scott Group dipping 25°SE. The latter consists of 1.5m of blocky grey and brownish grey silty mudstones weathering orange, with tightly packed joints, overlying a 8cm thick clay, weathering orange. This is a bentonite, a decayed (now clay-rich) volcanic ash. The bentonite overlies 12cm of grey mudstones before another 5cm bentonite is reached overlying 0.6m of grey blocky mudstones.

Bentonites are extremely useful for absolute (radiometric and fission track) dating, and the highest of these was dated by Ross *et al* (1982) at 466 million years old, with an error of ± 12 million years.

A little way back upstream there are exposures in blocky, irregularly bedded, brown-grey, micaceous and calcareous siltstones with calcareous nodules. About 1.5m are exposed and dip under the exposures of the main outcrop.

As you return upstream to the stile a small "beach" on the south side of the River Onny contains pebbles and boulders of numerous Ordovician, Silurian and Precambrian rock types which have been brought downstream.

Locality 8: View point on the south bank of the River Onny showing the famous Silurian/Ordovician unconformity.

Grid ref: SO 42638525

From locality 7 rejoin the field above the river by using the same stile as on the approach. Walk east for 300m along the field boundary above the river, crossing a stile

halfway, and from the south bank of the River Onny there is a superb view of the famous Silurian/Ordovician unconformity. This locality occupies a special place in the development of Lower Palaeozoic stratigraphy. It was here in the 19th century that the major unconformity between Murchison's Upper and Lower Silurian was first recognised, and this was one line of evidence which led to his Lower Silurian being newly termed the Ordovician System by Lapworth in 1879.

On no account cross the river to the north bank as this is outside the ownership of Cheney Longville Estates.

The exposure shows 3m of the highest Ordovician (Caradoc Series) rocks of this area, the Onny Shales, overlain unconformably by Silurian (Upper Llandovery) Hughley (Purple) Shales, of which approximately 3m can be seen.

The unconformity is not easy to pick out as the two rock formations both dip downstream (to the southeast) at similar angles. However, the Onny Shales comprise blocky blue-grey micaceous mudstones weathering orange, dipping SE at 22°. The highest 1m is more greyish green in colour in the centre of the exposure, probably because the rocks here are always above the river level and never under water. 10m further east where the unconformity dips below water level the highest Onny Shales are again blue-grey weathering orange. The unconformably overlying Hughley Shales consist of thinly bedded greenish brown, occasionally purple, shales and thin sandstones, dipping SE at 18°. Thin, pale creamy brown bentonites also occur just above the unconformity, particularly well seen in the winter when the cliff face is wet.

The Onny Shales are well known for trinucleid trilobites, in particular *Onnia superba*, and at times of low water rock samples can be obtained from the south bank of the river, and below water level, opposite and just upstream from the unconformity.

The unconformity is best identified by noticing the slight difference in dip between the two formations and also the fact that the Silurian beds show overlap to the west with each successive bed reaching further west than that below. Thus the angle of dip of the Hughley Shales is slightly less than the dip of the plane of the unconformity.

The whole Silurian sequence in this area oversteps (cuts out) the Ordovician sequence, resting on successively older and older formations to the northeast, eventually coming to rest on Precambrian rocks in places (see British Geological Survey 1:50000 map, sheet 166).

The unconformity has been caused by late Ordovician (Shelveian) earth movements which caused uplift, folding and faulting of the Ordovician rocks in this area during late Ordovician Ashgillian times. No rocks of this age occur in Shropshire (apart from in the far northwest, west of Oswestry).

These movements represent an important event in the development of the British Isles using a plate tectonic model (Toghill, 1990). It is thought that at this time southern Britain as part of the Avalonian microcontinent collided with a small continental area (Baltica), and this collision during the Ashgill, when southern Britain lay on the south side of the closing Iapetus Ocean at a latitude of around 30° south of the palaeo-equator, caused folding, faulting and uplift. Uplift, and possibly late Ordovician glaciation further south over North Africa, caused the sea to retreat westward from this area to a north-south shoreline between Oswestry and Welshpool during the Ashgill. During the succeeding Llandovery epoch of the early Silurian period the sea again spread (transgressed) from the west and eventually laid down the Hughley Shales (late Llandovery) over this part of Shropshire marked by an unconformity at their base.

From locality 7, which is the end of the trail, parties can retrace their steps, 1.75m west. Alternatively, a walk of 400m further east (downstream) leads to a footbridge over the River Onny, and a track leads north for 150m reaching the A489 at Crossway. (SO 429852).

Additional localities nearby

Immediately west of Crossway and towards Glenburrell, widening of the A489 has produced excellent exposures in the Cheney Longville Flags and Chatwall Sandstone. Here, the various fossils of the formations can be collected and the sedimentary structures studied. However, this is a very dangerous road, parking is difficult, and parties should take great care.

Wart Hill at SO 401847 is a steep hill of Precambrian Uriconian volcanics. It provides excellent views of the surrounding countryside and can be ascended easily from its south side via a minor road from north of Craven Arms to Edgton. However, the best viewpoint in the area for appreciating the Church Stretton Fault is looking north from the minor road leading from Cheney Longville to Edgton at SO 399853 just west of Upper Carwood. The view encompasses the southern Longmynd, the Church Stretton Fault Valley and the Uriconian Stretton Hills. Immediately to the NE is the incised valley cut by glacial melt waters along the line of the Church Stretton Fault and now followed by the B4370 road from Horderley to Marshbrook.

Further north the famous roadside unconformity at **Hope Bowdler** can be visited (SO 474924), where Harnage Shales rest on Uriconian

18

volcanics, and nearby the famous **Hazler Quarry** (SO 463924) shows Neptunean dykes (or fissures) in Uriconian volcanics with Harnagian fill. *Permission to enter the quarry must be obtained from the bungalow immediately next to the quarry on its north side.*

A quarry at SO 450895, west of **Acton Scott Church**, shows good exposures in the Acton Scott Limestone of the Acton Scott Group with characteristic fossils. The quarry is reached by a footpath between the church and Church Farm.

Permission must be obtained from Church Farm for access to the quarry.

ACKNOWLEDGEMENTS

Major S. W. Minton Beddoes has agreed to access for this trail on land which lies entirely within Cheney Longville Estates. It was also his suggestion for a more organised route for geological parties on his land which led to the writing of this guide and the setting out of the trail, and the author is grateful for his support. Those using the trail are asked not to divert away from the stated route.

Michael Harley from the Earth Science Branch of English Nature (previously the Nature Conservancy Council) and Andrew Hearle, English Nature's Conservation Officer for Shropshire, gave valuable advice and support. John Mycock, South Shropshire Countryside Project, arranged for stiles to be erected, and Martin Davies of the National Trust organised volunteers to clean up all the localities and provide steps at certain places.

The author is extremely grateful to Dr. Adrian Rushton of the British Geological Survey who provided up-to-date lists of the typical fossils of the various formations. Imogen Mortimer helped with the surveying of the various sites.

The geological map of Shropshire was redrawn by Colin Stuart of University College London.

REFERENCES

Maps
Ordnance Survey
Ludlow and Wenlock Edge – Sheet 137 (1:50,000)
Sheet SO 48 (1:25,000)
Geological Survey
Church Stretton – Sheet 166 (1:50,000)
Craven Arms – Sheet SO 48 (1:25,000)

Books
GRIEG, D.C., WRIGHT, J.E., HAINS, B.A., and MITCHELL, G.H. 1968. *Geology of the Country around Church Stretton, Craven Arms, Wenlock Edge and Brown Clee (Sheet 166)*. Memoir Brit. Geol. Survey H.M.S.O., London.
HAINS, B.A. 1969. *Geology of the Craven Arms area*. (Explanation of 1: 25000 Geological Sheet SO 48) HMSO, London.
TOGHILL, P. 1990. *Geology in Shropshire*. Swan Hill Press, Shrewsbury.
British Palaeozoic Fossils London. British Museum (Natural History).

Selected references from other publications.
BRENCHLEY, P.J. & NEWALL, G. 1982. Storm influenced inner-shelf sand lobes in the Caradoc (Ordovician) of Shropshire, England. *J. Sedim. Petrol.*, **52**, 1257-69.
DEAN, W.T. 1958. The faunal succession in the Caradoc Series of South Shropshire. *Bull. Br. Mus. Nat. Hist. (Geol.)*, **3**, 191-231.
1964. The geology of the Ordovician and adjacent strata in the southern Caradoc district of Shropshire. *Bull. Br. Mus. Nat. Hist. (Geol.)*, **9**, 257-96.
HURST, J.H. 1979a. The environment of deposition of the Caradoc *alternata* Limestone and contiguous deposits of Salop. *Geol. J.*, **14**, 14-50.
HURST, J.H. 1979b. The stratigraphy and brachiopods of the upper part of the type Caradoc of South Salop. *Bull. Br. Mus. Nat. Hist., (Geol.)*, **32**, 183-304.
ROSS, R.J. *et al.* 1982. Fission-track dating of British Ordovician and Silurian stratotypes. *Geol. Mag.*, **119**, 135-53.
WHITTARD, W.F. 1952. A geology of South Shropshire. *Proc. Geol. Ass.*, **63**, 143-97.
WHITTARD, W.F. 1958. *Geology of some classic British areas: Geological itineraries in South Shropshire*. Geologists' Association Guide, No. 27. Revised edition by DEAN, W.T. published 1968.